Work 99

像蜘蛛一样纺纱

Spinning like a Spider

Gunter Pauli

[比] 冈特·鲍利　著

[哥伦] 凯瑟琳娜·巴赫　绘

姚晨辉　译

上海远东出版社

丛书编委会

主　任: 田成川

副主任: 何家振　闫世东　林　玉

委　员: 李原原　翟致信　靳增江　史国鹏　梁雅丽
　　　　任泽林　陈　卫　薛　梅　王　岢　郑循如
　　　　彭　勇　王梦雨

特别感谢以下热心人士对童书工作的支持:

目录

Contents

一只蜘蛛正忙着织一张巨大的网，这张网至少有1平方米大小，而且肯定是这附近最大的一张网。

一只蝙蝠绕着网飞来飞去，评头论足道："这不仅仅是一张网，更是一件艺术品！"

"谢谢你的夸奖。"黄金圆蛛回应道，"这确实是一件独特的作品，需要用技巧和耐心来构建。在我开始织网之前，我就对这个位置进行了研究，并应用了数学原理。"

A spider is busy spinning a huge web. It is at least a metre by a metre in size, and must be one of the biggest in the neighbourhood.

A bat swings around and remarks, "This is not just a web; it is a piece of art!"

"Thank you for the compliment," responds the golden orb spider. "It is indeed a unique piece that requires skill and patience to build. Before I started spinning, I studied this site and applied the maths."

这不仅仅是一张网，更是一件艺术品！

This is not just a web; it is a piece of art!

即使是蛇也会被我的网缠住

Even snakes get entangled in my web

"你的网太大了，不会只捕捉到昆虫。你是否偶尔也用它抓到过鸟呢？"

"即使是蛇也会被我的网缠住。有一次，一条小蛇被我的网困住，最后把我的网弄破了。当然，我不敢吃她，只是把网修补好了。"

"既然你吐的丝那么强韧，人类难道不会对你感兴趣吗？"

"𝒴our web is too big to only catch insects. Don't you catch birds in it once in a while as well?"

"𝐸ven snakes get entangled in my web. Once, a young one ended up in my web and destroyed it. Of course I did not dare to eat her; I only recovered my silk."

"𝒮eeing that your silk is so strong, have people not become interested in you?"

"当然会！但我们
蜘蛛不喜欢规则和约束，因此
很难被驯化。"
"那人类有没有试图收取你的网？"
"哦，有的，一些人试图用我的丝做衣服，他们甚
至尝试用我的丝织成手套、降落伞、渔网
和绳索。"

"You bet! We spiders are quite
unruly and undisciplined however, so it is
difficult to domesticate us."
"Have people tried to harvest your webs yet?"
"Oh yes, some tried to make clothing out of
my silk and they even tried to turn my fibres
into gloves, parachutes, fishing nets, and
ropes."

我们蜘蛛不喜欢规则和约束

We spiders are quite unruly and undisciplined

...our silk has been used in Asia as bandages

"我听说你甚至可以入药。"

"确实是这样。多年来，我们的丝在亚洲被当作绷带，用来止血。现在，英国的一些聪明人正在研究如何使用我们的丝培育神经。"

"培育神经？"

"I heard that you're even getting into medicine."

"Yes, indeed. For years our silk has been used in Asia as bandages to stop bleeding. But now some smart people in England figured out how to use our silk to grow nerves."

"Grow nerves?"

"是啊！还在培育肌肉和
骨骼之间的软骨！"
"培育神经和组织……这听起来几乎太不可思
议了！"
"听起来是这样，但我的丝确实有助于细胞生长，
这是一个可以分享的美好礼物。"

"Yes! And to grow cartilage
between muscles and bones!"
"Grow nerves and tissue ...
that sounds almost too magical."
"It is, but my silk really does help cells to
grow. It is a wonderful gift to be able to
share."

......培育肌肉和骨骼之间的软骨！

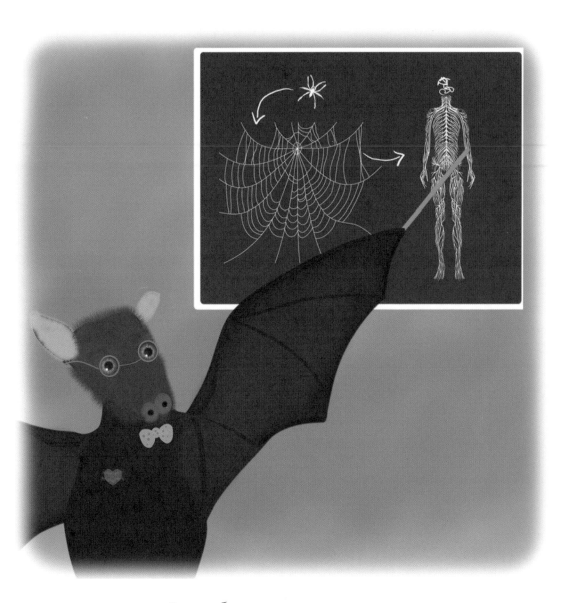

...grow cartilage between muscles and bones!

······用你的丝供应全世界?

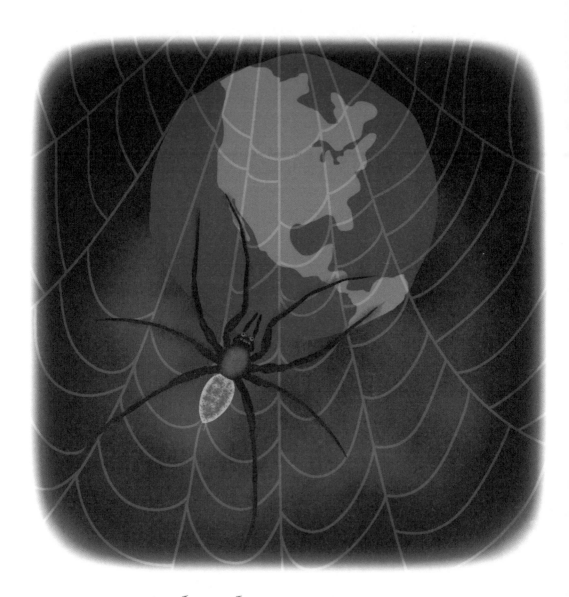

...to supply the whole world with your silk?

"这会让你非常
受欢迎的，不过你如此精美
的丝能够供应全世界吗？"
"想要满足全世界对丝的需求，黄金圆蛛的数量永
远也不够多。"
"所以说这只不过是一个画饼充饥的梦想？"蝙蝠问
道，"你最终会让太多的人失望的。"

"This will make you very popular, but
will you ever be able to supply the whole
world with your wonderful silk?"
"There will never be enough golden orb spiders
to make all the silk the world needs."
"So is this not just a big pie-in-the-sky
idea then?" asks the bat. "You will end up
disappointing so many people."

"你为什么会这么说？我已经与蚕达成了协议，让他们为我的项目提供帮助。"

"他们是食草动物，你是食肉动物。你确定你们的丝是一样的吗？"

"Why do you say that?
I have actually made a deal with the
silkworms to help me with my project."
"They are herbivores; you are a carnivore.
Are you sure you make the same silk?"

……与蚕达成了协议

...made a deal with the silkworms

......我们可以将其转换为一种更强韧的材料

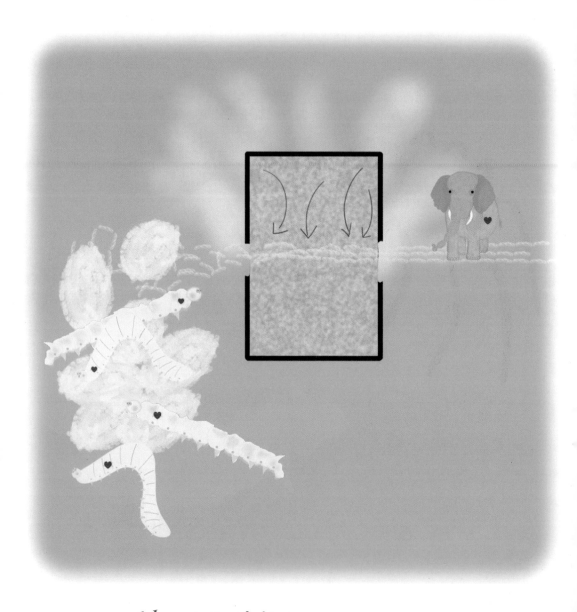

...we could convert it into a stronger material

"不，虽然我们都吐丝，但性质并不相同。然而，如果我们给柔软的丝增加压力和水分，我们可以将其转换为一种更强韧的材料。"

"有多强韧？"

"这么说吧，也许强韧到有一天你可以利用我的丝再造人体的组成部分。"

"Yes, we both make silk, but not the same kind. If we, however, add pressure and moisture to their soft silk, we could convert it into a stronger material."

"How strong?"

"Well, maybe so strong that one day you could regenerate parts of the human body from my silk."

"你的丝真的很特别！"

"没错，但是要想一个新的点子获得成功，最好能够改变游戏规则。"

"所以你没有改变和破坏任何规则，而是利用科学的奇迹创造了新的规则？"

"是的，我想用这种方法让这个世界变得更美好。"

……这仅仅是开始！……

"You silk is very special indeed!"

"That's true, but if you want to be successful with a new idea, you'd better change the rules of the game."

"So you are not bending and breaking any rules, you are making new ones by using the marvels of science?"

"Exactly. And that is how I intend to make this a better world."

... AND IT HAS ONLY JUST BEGUN!...

...... 这仅仅是开始！ ......

... AND IT HAS ONLY JUST BEGUN! ..

The web of the golden orb spider could be called a "web of steel". Measuring up to one-by-one metre, it is not only the strongest and largest kind of web, but can last for several days. The silk boasts a golden colour.

黄金圆蛛的网可以被称为"钢丝网"。根据对1平方米蛛网的测量，它不仅是一种最强韧和面积最大的网，而且可以保持数天之久。黄金圆蛛的丝呈现出一种耀眼的金色。

The web of the golden orb spider can trap small birds, which the spider does not eat. To avoid damage, the spider leaves silk curtains on its web, which act like a safety strip across a glass door and builds strong barrage threads around the main web.

黄金圆蛛可以用网捕获它并不吃的小型鸟类。为避免蛛网被破坏，蜘蛛在网上放置了丝帘，其作用类似于玻璃门上的安全条，围绕主网构建强韧的拦阻线。

雄性黄金圆蛛的大小约是雌性黄金圆蛛的千分之一，它住在网上，窃取雌性黄金圆蛛的食物。

The male golden orb spider is about 1,000 times smaller than the female. He lives in the web, and steals her food.

蜘蛛丝比合成芳纶坚韧得多，合成芳纶的生产需要刺激性的化学物质。蜘蛛丝是由蛋白质制成的，可以回收，而且蜘蛛能够当场对蛛网随意翻新。

Spider silk is much tougher than the synthetic Kevlar, which requires harsh chemicals for its production. Spider silk is made from protein and can be recycled and reconditioned indefinitely on site by the spider.

The Madagascar golden orb spider has very poor vision. Instead of using sight, it uses its very sensitive sense of touch to feel vibrations on the web and track down its prey.

马达加斯加黄金圆蛛的视力很差，因此它不是直接用眼睛看，而是利用非常灵敏的触觉感受蛛网的振动并追踪猎物。

Some spiders attract their prey with the odour of dead, rotting organic matter blended into their webs. Some spiders also use chemical attractants in their silk. A spider either eats its meal right away or wraps up its prey in silk and stores it for a time when food is scarce.

一些蜘蛛将死去的、腐烂的有机物融入网中，利用它们散发的气味吸引猎物。也有一些蜘蛛在它们的网中使用一些引诱性的化学物质。捕获猎物后，蜘蛛要么立即将猎物吃掉，要么用蛛丝将猎物包裹并储存起来，等食物短缺时再吃。

蜘蛛织网用的建筑材料是从自己的身体中制造出来的。试想一下，如果我们可以利用自己的细胞建造住宅那会怎样？蜘蛛会吃掉全部或部分旧网，再用可回收翻新的材料来编织新网或蛛网的一部分。

Spiders produce the building materials for their webs from their own bodies. Imagine if we could construct dwellings from our own cells! Spiders consume all or parts of their old webs and then weave new webs or parts of their webs with recycled and reconditioned material

蛛丝，或者与其类似的蚕丝，是具有生物相容性的，可被用作软骨的替代物，促进细胞增殖，并可作为神经再生的引导材料。

Spider silk, or its analogue produced from silkworms' silk, is biocompatible, can be used as a substitute for cartilage, promote cell proliferation, and act as guiding material for nerve regrowth.

Would you farm with spiders to produce silk? Or do you have a better idea?

你会养殖蜘蛛生产蛛丝吗？你有没有更好的想法？

你能想象用蛛丝来移植头发吗？如果用蛛丝做原料，如何来移植头发？

Can you imagine removing hair with silk? If silk is the raw material, how will hairs be removed?

Are you planning to improve yourself with existing rules, or will you become the best you can be by making new rules?

你是否打算在现有规则下提升自己，还是说通过制定新的规则，你就能做到最好？

你认为那些不可能实现的梦想，最终会让起初欣赏这一想法的人失望吗？

Do you think dreams that are impossible to achieve will ultimately disappoint the people who loved an idea at the outset?

Do It Yourself!

自己动手!

Go outside and look for a spider in the garden. It will not be hard to find. Do not touch the spider with your bare hands unless your guardian is 100% sure it is not venomous and will not bite you. Try to find its web and look at the construction of the web. Find the outer corners, look where the sun is, figure out where the wind comes from, and check both sides (front and back) of the web. Assess the logic the spider must have used to build the web in that specific place. Does it make sense, and why?

到室外花园里寻找一只蜘蛛。这并不难。但是不要直接用手触碰蜘蛛，除非你的父母完全肯定它是无毒的，而且不会咬你。试着找到它的网，观察一下网的构造。找到转角的位置，看看太阳在什么地方，确定一下风向，并检查蛛网的两边（前面和后面）。推测一下蜘蛛在这个特定位置织网肯定会用到的某种逻辑，在这里织网有道理吗？为什么？

学科知识
Academic Knowledge

生物学	蜘蛛是呼吸空气的节肢动物，拥有40 000多个不同品种；蜘蛛是食肉性动物；在某些种类中，成千上万的蜘蛛会共用蛛网；蜘蛛一般只有不到一年的寿命，那些长寿的蜘蛛会用丝带编织越冬的巢，帮助它们抵御寒冬；蛛丝具有避免细菌和真菌感染的作用；软骨是依附在骨头上的柔性结缔组织，没有骨头那么硬和脆，但比肌肉更硬且弹性较差。
化学	蛛丝由蛋白质组成；醌类物质可能是导致黄金圆蛛的网呈现金色的原因；对蜘蛛毒素的研究导致了β受体阻滞剂的发现。
物理	阳光照射下的蛛网能诱捕蜜蜂，因为蜜蜂会受到模拟蓝天的紫外反射区的吸引；在阴凉处，蛛网的黄色融入了背景树叶的颜色之中，充当伪装；蜘蛛会拆除其蛛网较低的部分，确保在强风吹过时蛛网不会被破坏；结构化的结晶丝向非结晶丝的转换。
工程学	利用压力和湿度对毛虫（蚕）吐出的丝进行转化，使其与黄金圆蛛的丝具有相同功能；蜘蛛使用步足将丝从身体中抽出来；为了加快纺丝过程，蜘蛛会迎着风，利用风的力量来加速；芳纶具有阻挡子弹的能力。
经济学	对蜘蛛网的工作方式可以进行成本效益分析：如果没有功能性的蛛网，蜘蛛会饿死，如果蜘蛛投资不足（编织了一个小网），蛛网将无法获得足够的食物，如果蜘蛛投资过多，将导致其死亡。
伦理学	从学习关于自然的知识到向自然学习的转换。
历史	古埃及、希腊、伊斯兰、日本和非洲的神话，印第安文化以及澳大利亚土著艺术中都有以蜘蛛为主角的故事；术语"万维网"让我们联想到蜘蛛网。
地理	新几内亚部落将蜘蛛视为一种美好的东西，而不是威胁；南太平洋岛屿上使用蛛网制作渔栅和渔网；所罗门群岛上的人用蛛网做的球来困住鲨鱼；制作一件披肩和斗篷需要耗费超过一百万只马达加斯加黄金圆蛛的蛛丝。
数学	蜘蛛编织圆形的网，利用不同的锚点创建复杂的图案；蛛网理论或蜘蛛的数学：蜘蛛只用直线来创建圆形图案，编织切线而不是弧线；蜘蛛利用其遗传本能创造了一个数学奇迹。
生活方式	蜘蛛非常注意节约材料，它使用自己身体产生的物质来建造蛛网，并对旧的材料进行回收。
社会学	我们可以将自然界中物理结构的创造视为"艺术"，或者说它鼓励人们艺术性地表达自己。
心理学	感知取决于观点：你是一个侏儒还是一个巨人？自然界是一个创造性表达的场所；蜘蛛恐惧症：对蜘蛛的病理性恐惧。
系统论	蜘蛛已经进化到"理解"生态系统的背景，并加以利用为自己谋利，它优化了材料的使用；蜘蛛和人类之间的关系表明了相互关联性：蜘蛛为人类创造了一个生产再生医药的机会。

情感智慧
Emotional Intelligence

蝙 蝠　　　　蝙蝠通过赞美蜘蛛开始了谈话。他观察蛛网并注意到网的大小。他抛出一系列对蜘蛛有吸引力的问题将谈话继续下去。蝙蝠见多识广，准备充分。尽管也有自己的神奇事迹可供吹嘘，他仍保持着谦虚的态度，赞美蜘蛛的独特性。他鼓励蜘蛛对所有人大胆说出其成就。蝙蝠担心是否有足够的蛛丝提供给大家，温和地告诫蜘蛛不要让大家失望。他通过询问一系列问题，了解到更详细的情况。最后，蝙蝠就蜘蛛对生态系统的重要性，以及如果具有类似蜘蛛的能力可能带来何种社会变革发表了哲学性的评论，结束了这次谈话。

蜘 蛛　　　　面对赞美，蜘蛛受宠若惊。她充满自信，毫无保留地共享了信息。通过叙述奇闻轶事，蜘蛛在她和她的蝙蝠朋友之间营造了轻松的氛围。她意识到了蜘蛛工作的价值和重要性是由人们表现出来的兴趣所驱动的，提供了各种可能性的细节情况。蜘蛛具有战略方针和预见性。她没有感到任何压力，坚信自己能够响应大家的需求，不会让他们感到失望。她认为通过帮助发明新的医疗方法和创造新的就业机会，她可以对环境作出重要贡献。这是一个关于世界是如何联系在一起，以及个体的行为如何让他人的生活和环境更美好的典范。然而，成功并不仅仅取决于技术性能，还依赖于个体为创造一个更美好的世界而改变游戏规则的能力。

艺术
The Arts

　　让我们来制作蜘蛛和蝙蝠纸链！将纸折叠，并剪成蜘蛛或飞行蝙蝠的形状，悬挂起来作为装饰。可上网搜索制作方法。

思维拓展
Systems: Making the Connections

蜘蛛是一个非凡的物种，它们的吐丝和织网技术性能值得赞赏。然而，蜘蛛真正突出的特点是它们的资源利用效率。这归功于它们的饮食习惯，它们的身体能够产生自己生存所需的所有材料。蜘蛛可以回收所有的材料，并利用它们的旧网再生新的蛛丝，来编织一个新网捕捉食物，或者建造一个庇护所以便度过寒冷的冬天。蜘蛛会根据场地来调整自己蛛网的建设，并会考虑到太阳的运动、月亮的位置、树冠的保护、鸟类的飞行路径以及风向等因素。看起来蜘蛛已经塑造了它们独特的生态定位，并通过侵入数千个生态位来展示它们的资源效率和适应能力。由于这种能力，目前在地球上有超过40 000种蜘蛛，而且这个数字还在不断增长。蜘蛛以一种独特的方式运用数学和几何学理论编织它们的网，它们只使用干净的水作为溶剂，它们的丝由蛋白纤维组成，完全可回收、可再生且适应性强。蜘蛛让我们深入了解作为社会系统或生态系统的一部分，如何让自己适应自然，以确保可持续性以及应对自己的需要，同时也有利于作为系统一部分的其他物种（如人类）。

动手能力
Capacity to Implement

列出蜘蛛网所有潜在的商业应用。首先看一下技术，列出那些以受蜘蛛启发的创新作为核心商业模式的公司。然后看一下蜘蛛的工作方法——它吐丝和织网的方式。它遵循的执行原则是什么？看看它如何进行回收利用，如何通过观察风向和伪装自己来适应环境。这会让你深入了解一个行业如何通过回收利用和自适应等可持续的做法，使自己成为有机化的结构。此外还有很多。蜘蛛不仅启示我们如何解决当今的挑战，同时也向我们展示了如何通过改变游戏规则建立新的商业模式来获得成功。试着找出游戏的规则是什么——从产品到加工、到系统。

故事灵感来自
This Fable Is Inspired by

弗里茨·沃尔拉特
Fritz Vollrath

弗里茨·沃尔拉特在德国弗赖堡大学动物学系获得博士学位。他在巴拿马史密森尼热带研究中心工作时，发现了黄金圆蛛的能力。他被这种蜘蛛的表现深深打动。他研究了蛛网的工程学技巧，得知人们可以将商用的、大规模生产的毛虫（蚕）的丝转化为类似蜘蛛丝的工程材料。根据他的研究，已经成立了致力于生物材料、医用缝合线、神经再生和软骨替换技术的四家公司。他发自内心地喜爱大自然，喜欢所有的动物，从大动物（大象）到小动物（蜘蛛）。他是拯救大象基金会的主席，每年花费大量的时间在肯尼亚的帕拉研究中心研究大象以及它们的社会和生态系统。

图书在版编目（CIP）数据

冈特生态童书.第三辑修订版：全36册：汉英对照 /
（比）冈特·鲍利著；（哥伦）凯瑟琳娜·巴赫绘；
何家振等译.—上海：上海远东出版社,2022
书名原文：Gunter's Fables
ISBN 978-7-5476-1850-9

Ⅰ.①冈… Ⅱ.①冈… ②凯… ③何… Ⅲ.①生态环
境−环境保护−儿童读物—汉、英 Ⅳ.①X171.1−49

中国版本图书馆CIP数据核字（2022）第163904号
著作权合同登记号图字09−2022−0637号

策　　划　张　蓉
责任编辑　程云琦
封面设计　魏　来李　廉

冈特生态童书

像蜘蛛一样纺纱

[比]冈特·鲍利　著
[哥伦]凯瑟琳娜·巴赫　绘
姚晨辉　译

记得要和身边的小朋友分享环保知识哦！
八喜冰淇淋祝你成为环保小使者！